Chinese Cities
Beijing Impressions

中国城市
北京印象

五洲传播出版社
China Intercontinental Press

Revisiting the Past

访古　古都故事

Palace Museum
Imperial Palace of the Ming and Qing Dynasties

故宫 | 穿 梭 时 光 的 故 城

The Palace Museum was the imperial palace of the Ming and Qing dynasties (1368-1911), well demonstrating the ancient wisdom of China. Walking across its vast space, one can be easily transported back over a history of 600 years.

In Spring, the tender green willows flanking the moat set off the exquisite turrets of the massive palace walls. The Forbidden City, a scene of great tranquility in the past, is now an attraction crowded with visitors, but its inherent imperial poise still remains.

　　故宫，融合着中国古老智慧的建筑。其一砖一瓦一砾仿佛都在述说明清两朝的帝王生活。穿梭其中，每一步仿佛都踏着六百年的故事和六百年的历史。

　　每到春日，护城河两岸的杨柳舒展新芽，点缀着精巧的角楼。昔日宁静的紫禁城，如今游人如织，但那与生俱来的帝王之气却依旧不改。

Summer Palace
Royal Garden Museum

颐和园 | 皇家园林博物馆

The Summer Palace is one of the best-preserved royal gardens in the world containing a wide variety of scenic spots. Willows dance in the breeze by the tranquil Kunming Lake forming a delightful poetic picture. The Heralding Spring Pavilion, Phoenix Mound and Qingyan Stone Boat add great luster to the scene. The gorgeous and complicated corridor that extends for over 1,000 meters always makes people feel they are part of this beautiful picture.

颐和园是世界上造景丰富、保存最完整的皇家园林之一。昆明湖边，杨柳拂面、晓风残月的诗情油然而生；知春亭、凤凰墩、清晏舫又使这诗情多了几分画意；绚丽繁复的千米长廊总让人不觉是人在画中，还是画在人间。

Temple of Heaven
Offering Sacrifices to Heaven

天坛 | 神 圣 的 祭 天 台 ————————————

As one of the architectural symbols in Beijing, the Temple of Heaven was included in UNESCO's *World Heritage List* in 1998. Old trees with coiled branches can be found everywhere in the park, telling the mysterious stories of the temple through the ages. It is advisable to walk leisurely on the path shaded by pines and cypresses where one can hear the melodious tones of the *erhu* (two-stringed musical instrument). The style of Beijing finds full expression there.

天坛，北京的象征之一，1998 年被联合国教科文组织列入"世界遗产名录"。天坛公园里遍植虬枝盘结的老树，向今人默默低语着昔日天坛的神奇。闲来无事时，若在松柏成林的小道上走一走，听一曲悠扬的二胡声，霎时间，北京的韵味便扑面而来。

Great Wall

Inside and Outside Scenes

长城 │ 城 墙 内 外

A Chinese saying has it that "one who fails to reach the Great Wall is not a hero". No other structure in China can equal the Great Wall in representing the spirit and verve of the Chinese people. As a transnational defense project, the Great Wall passes through a variety of landforms, including mountains, deserts, grassland and rivers. It is really a wonder in the history of ancient architectural engineering.

中国有句俗语"不到长城非好汉"。在中国，没有任何一处建筑可以比万里长城更能代表中国人的精神和气魄。作为世界性的防御工程，长城翻山越岭，穿沙漠、过草原、越绝壁、跨河流，所经之处地形复杂，不仅是中国建筑工程史上的奇迹，也是世界古代建筑工程史上的奇观。

Yonghegong Lamasery
Magnificent Royal Temple

雍和宫 | 华 贵 的 皇 家 寺 院 ────────

    The resplendent, magnificent and solemn Yonghegong Lamasery can be seen from the North Second Ring Road in Beijing. Its magnificence is shown from the main gate, followed by a long covered corridor for people to shake off the dust of modern life before entering the lotus-shaped gate.

　　站在北京北二环上望去，总会被雍和宫金碧辉煌和庄严肃穆的气势所震撼。那恢弘之气从山门便开始了，穿过那条遮天蔽日的甬道，仿佛在脚步的一起一落中抖落身上的尘土之后，才可以进入那道恰如盛开莲花的寺门。

Old Summer Palace
Vanishing Versailles

圆明园 ｜ 消 逝 的 凡 尔 塞

The old Summer Palace is a legendary royal garden, with a land area equivalent to the Forbidden City and a water area equaling that of the Summer Palace. It was plundered and burned down by British and French forces in the mid-19th Century. After decades of renovation, the scars can still be seen.

昔日的圆明园是中国皇家园林的一个传奇。圆明园的陆上建筑面积等同于故宫，而水域面积又等同于颐和园。1860 年，圆明园惨遭英法联军野蛮劫掠焚毁，虽经过数十年的整修，却依然难掩满目的伤痕。

Beihai Park and White Buddhist Pagoda
Picturesque Royal Garden

北海与白塔 | 图 画 中 的 皇 家 园 林

After ceaseless construction during various dynasties, Beihai Park took shape. Having a passion for Jiangnan (area south of the Yangtze River), Emperor Qianlong created the many wonders still to be found in the park. The famous traveler Marco Polo once marveled at its beauty in his travelogue.

若说历代王朝不停地建造奠定了北海公园的初型，那酷爱江南的乾隆皇帝则倾心创造了北海的神奇，就连曾经走过千山万水的马可·波罗也曾在游记里感慨过它的美丽。

Peking University

School of Higher Learning by Weiming Lake

北大 ｜ 未 名 湖 畔 自 由 的 殿 堂

As a center of the new culture movement, Peking University has left an indelible page in the modern history of China. Standing on Boya Tower, one can overlook Weiming Lake surrounded by weeping willows that are much admired by every passerby.

北京大学作为新文化运动的中心，在中国近代史上留下不可磨灭的一页。博雅塔和未名湖是北大的灵性所在，博雅塔温文尔雅地俯视着湖面，湖柳垂首梳妆摇曳顾盼，路人的脚步到了这里也会不由地慢下来。

Tsinghua University
Moonlit Lotus Pond in the Most Beautiful School

清华 | 最 美 学 堂 里 的 荷 塘 月 色

As the first building of Tsinghua Campus, the Gongzi Hall has witnessed the vicissitudes and hard-working course of the university. The most charming campus scene is the Moonlight over the Lotus Pond described by Zhu Ziqing (1898-1948). Nowadays, the lotus flowers fill the pond embraced by a hill and trees.

最初的清华园是从工字厅开始，那里见证了清华的百年沧桑和自强不息的历程。清华园内最令人神往的地方当属朱自清笔下的荷塘月色了，如今开满池塘的荷花在山丘和树林地掩映下生机勃勃。

Nanluogu Lane and Wudaoying
Pioneer of Beijing Arts

南锣鼓巷与五道营 | 文 艺 青 年 的 下 午

Nanluogu Lane was built in the Yuan Dynasty and is one of the oldest neighborhoods in Beijing. It is also the largest and best-preserved traditional A *hutong* lane in Beijing is full of literary and artistic creativity. Numerous creative shops are lined up in a distance of 600 meters against a background of interesting plants and pleasant furnishings.

南锣鼓巷是北京最古老的街区之一，是完整保存着元代胡同院落肌理、规模最大、品级最高、资源最丰富的棋盘式传统民居区。这里分布着大量的古代民间建筑和名人故居。和南锣同样文艺的胡同就数五道营胡同了，这条 600 米的胡同布满了极具创意的小店，也有着情趣的绿植及居家般怡然自得的氛围。

798 Art Zone

Art Avant-garde in Beijing

798艺术街区 ｜ 北京的艺术先锋派

The 798 Art Zone was rebuilt on the site of a waste ordnance factory. It is full of bold sculptures, avant-garde studios and post-modern bars.

798 是由北京废弃的军工厂改造而成的时尚艺术街区。整个街区布满了大胆的雕塑、前卫的画室和充满后现代感的酒吧。

Bird's Nest and Water Cube
New Buildings in Beijing

鸟巢与水立方 | 北 京 新 建 筑

The hollowing out technique, ceramic texture and warm red made the main stadium of the 2008 Beijing Olympic Games novel and avant-garde. In the shape of a bird's nest, it was thus named The Bird's Nest. In 2007, it was rated as one of the world's top 10 architectural wonders by *Time* magazine.

Beside the Bird's Nest, a blue building like a Crystal Palace is the Water Cube, the main swimming pool used for the 2008 Beijing Olympics. The two constructions glitter along the extension line of the central axis of Beijing.

镂空的手法、陶瓷的纹路、热烈的红色使 2008 年北京奥运会的主体育场呈现出一种新奇而前卫的效果，因为外观形似鸟窝，故以"鸟巢"命名。鸟巢在 2007 年被《时代》周刊评为世界十大建筑奇迹。

在鸟巢旁边如水晶宫般蓝盈盈的建筑是水立方，是北京奥运会的主游泳馆。鸟巢和水立方一起，在北京中轴线的延长线上熠熠生辉。

Sanlitun

Fashionable Sleepless Town

三里屯 | 时 尚 的 不 夜 城

Fashionable Sanlitun is a great place to experience Beijing's nightlife. There are countless bars, cafes and gourmet shops. Every day, many foreigners come to Sanlitun to experience Beijing's fashion and tolerance.

时尚的三里屯是体验北京夜生活的好去处，这里聚集着数不清的酒吧、咖啡馆、美食店，每天都有许多外国人来到三里屯感受北京的时尚与包容。

Beijing CBD
All Beijing Looks up

北京CBD | 全北京向上看

The Beijing CBD, with row upon row of skyscrapers, is the most populous, economically developed and fashionable area in Beijing. This unique area demonstrates the nature of Beijing as a modern international metropolis.

北京CBD里的摩天大楼鳞次栉比,是北京人口最多、经济最发达、最能够代表时尚潮流的区域。这片独具特色的区域展现出北京作为现代国际大都市的风貌。

Tradition

传统 历史传承

Imperial College

Highest Institution in Olden Time

国子监 | 昔日的最高学府

The Imperial College was the highest institution of higher learning in the Yuan, Ming and Qing dynasties (1271-1911). Each black brick in the courtyard and each cypress therein have witnessed the hard work of students of numerous generations. On sunny days, there are always people reading silently in the courtyard, continuing the great college tradition.

作为元、明、清三代最高学府的国子监，庭院内每一块青砖、每一棵柏树都见证着昔日学子的艰辛。如今在阳光正好的时节，人们总能看见庭院内静静读书的人，这画面正是国子监百年风貌的延续。

Quadrangle Dwellings
Memory of the Canopy and the Pomegranate Trees

四合院 | 天棚 和 石榴树 的 记忆

Such words as "sunshade, fish jar, pomegranates, *laoye* lord, fat dog and fat girl" were used to describe the life of wealthy old Beijingers. The old Beijingers who lived in the quadrangles in those years spent their time in a leisurely fashion. Today, the quadrangles still remain as a quiet, hidden presence amid the *hutongs*, retaining memories of a precious life.

　　"天棚、鱼缸、石榴树，老爷、肥狗、胖丫头"人们曾经用这些词语来描绘富裕的老北京人的生活。当年生活在四合院里的老北京人就是在这样宽敞而悠闲的岁月里打发着时光。如今，四合院依然静静地隐藏在胡同中，保留着那些珍贵的生活记忆。

Hutong
The Folk Life

胡同 | 市 井 生 活

There is a saying that Hutong represents the soul of Beijing. Hutong dates back the Yuan Dynasty, when the rulers used this system to divide and plan the city. The word "hutong" is a Mongolian term, and because houses were built alongside the river or near the well, it actually means "water well".

Take a stroll in hutong in the afternoon, you can always find siheyuan of classical style, where you will see elderly people sitting at the front door and basking in the sunshine. If there happens to be sound of music and singing in the air, that must be the community activity for the elderly. Stand still and listen to it for a while will make a good escape from the busy life in the modern city.

有人说，只有胡同才是北京人的魂。胡同原是一种城市规划形式，创始于元代。过去人们建造房屋大都依靠在水边或水井边，胡同一词翻译成为汉语就是"井"的意思。

在北京的胡同里，依然居住着那些老北京人。找个下午，到胡同里转转，看一看那古香古色的四合院门和晒太阳的老人。有时候，社区居委会里会有琴声咿咿呀呀响起，站在那里听一听，就是一种闹市中突然得来的享受。

Lao She Teahouse
Leisure Life over a Cup of Tea

老舍茶馆 | 一 杯 茶 里 的 闲 散 日 子

The Lao She Teahouse at Qianmen has always been booming. The waiters in robe welcome customers to the lobby. Sipping potted tea while listening to a cross talk performances, the customers can spend an interesting afternoon.

　　位于前门的老舍茶馆向来生意兴隆,穿长袍的伙计一声嘹亮高亢的"来啦您呐"便将客官迎进大堂。在茶馆里沏上一壶茶,再听上一段相声,客官们便可以在吆喝声、笑声、喝茶倒水声中度过一个有滋有味的下午。

Peking Opera
Drama Life on Stage

京剧 | 舞 台 上 的 戏 剧 人 生

Peking Opera is the quintessence of China with a history of more than 200 years. Going to the opera house was a favorite pastime of old Beijingers. Nowadays, there are fewer and fewer theaters in the capital. It is highly recommended to visit them and listen to the moving voices reflecting centuries of cultural development.

京剧是中国的国粹，流传至今已有 200 多年的历史。听戏是老北京人最喜欢的消遣。如今的戏楼越来越少，空闲下来，我们可以去这些戏园子坐坐，听一听那些咿咿呀呀的动人声音。

Crosstalk
Tianqiao Memory

相声 | 天 桥 记 忆 ————————————

Enjoying a crosstalk show at the Tianqiao Teahouse was a tradition of the old Beijingers. With various bars and drama performances, Tianqiao made visitors unwilling to go back home. Nowadays, crosstalk performances are still a feature in many teahouses and theatres. The actors on the stage are speaking, imitating, teasing and singing, winning applause from the audience. Crosstalk performances are increasingly sought after by today's young people.

到天桥茶馆听相声是老北京人的传统，"酒旗戏鼓天桥市，多少游人不忆家"就是对北京老天桥昔日生活的描述。如今很多茶馆和戏院里依然保留着相声节目，台上的演员一阵说学逗唱，台下的观众们便是一阵吆喝鼓掌。如今的相声表演也越来越受年轻人的追捧。

Sugar-coated Haws
Childhood Memory in Winter

冰糖葫芦 | 冬 天 里 的 儿 时 记 忆

The winter in Beijing is chilly, yet warm. Especially after the snow, the city seems to be covered with a layer of fine white sugar. The cries of Crispy Sugar-Coated Fruit sellers in the hutongs of Beijing are still loud and clear in Beijingers' childhood memories.

在北京人的童年记忆中，北京的冬天在寒意中透着温暖，尤其是在雪后，满城像蒙着一层细细的白砂糖，胡同里冰糖葫芦的叫卖声声声入耳，悠远清长。

Instant-boiled Mutton

Must-taste Food

涮羊肉 | 锅 里 有 乾 坤

On a street of Beijing in the winter, one sees a restaurant with a thick layer of steam misting up its glass window, through which one can just discern a copper pot on a windowsill - definitely a hot pot restaurant! Sitting around a hot pot, the diners talk and laugh warmly, occasionally having a look at the falling snow outside the window. This is the way to defy the rigors of winter.

冬日北京的街头，如果你看见一家餐馆的玻璃上糊着一层厚厚的蒸汽，窗口处还摆放着一个紫铜锅子，那么这必是一家涮肉馆无疑。人们围坐在一口锅子四周，谈笑风生、暖意融融，偶尔看看窗外飘落的雪花，不失为冬天最值得期待的日子。

Beijing Roast Duck
Century-old Food

北京烤鸭 | 百 年 美 食 传 奇

Originally roasted in a closed oven in the imperial palace of the Ming dynasty (1368-1644), Beijing Roast Duck was introduced when the capital was moved from Nanjing to Beijing. The roast duck slices are dipped in a sweetened sauce and then wrapped with shredded onion in lotus leaf shaped pancakes, a famous eating method. Foreigners see "visiting the Great Wall and having roast duck" as two musts during a visit to Beijing.

北京烤鸭是从明代宫廷的焖炉烤鸭随着迁都传到北京的。这种烤鸭以用荷叶饼蘸裹葱丝、甜面酱的吃法闻名天下，外国人都把"游长城，吃烤鸭"作为来北京所必须做的两件事。

Old Beijing Snacks
Unique to the Imperial Town

老北京小吃 | 老 城 里 的 烟 火 气

Early in the morning, in a *hutong* in Beijing, one can always see a steaming snack bar, which offers "fried liver", steamed buns, fried cakes, and small wonton, making people drool with anticipation.

When the evening lights are lit, the time-honored shops in Beijing hang lanterns outside, and the waiters in robes welcome customers in their own unique way. Sitting in an old-fashioned wooden armchair by an Eight Immortals Table, one can have a celadon pot of tea and a bowl of fried bean-paste noodles with eight small side dishes. This is a scene of the tranquil life to which Beijing people like to cling.

大清早在北京的胡同里逛一逛，总能遇见冒着水汽的小吃店，里面有炒肝、包子、炸油饼、小馄饨，让人垂涎欲滴。

华灯初上的时候，北京老字号铺子外挂着灯笼，屋里灯火通明，穿堂的伙计正吆喝着"您里边儿请"。八仙桌太师椅，青瓷大茶壶沏一碗盖碗茶，再来一碗八小盘菜码的炸酱面，北京人的日子就这样热气腾腾地过出了滋味。

图书在版编目（ＣＩＰ）数据

北京印象：汉英对照 / 达雅著；王国振译 . -- 北
京：五洲传播出版社 , 2020.1（2023.5 重印）
（中国城市）
ISBN 978-7-5085-4336-9

Ⅰ.①北… Ⅱ.①达… ②王… Ⅲ.①本册②北京—
概况—汉、英 Ⅳ.① TS951.5 ② K291

中国版本图书馆 CIP 数据核字 (2019) 第 268706 号

中国城市：北京印象
Chinese Cities：Beijing Impressions

出 版 人 ： 关　宏
责任编辑： 杨　雪
设计策划： 理想集
　　　　　 287201901
插　　画： 王建华
文　　字： 达　雅
译　　者： 王国振
装　　帧： 张伯阳
出版发行： 五洲传播出版社
地　　址： 北京市海淀区北三环中路 31 号生产力大楼 B 座 6 层
邮　　编： 100088
发行电话： 010-82005927，010-82007837
网　　址： http://www.cicc.org.cn，http://www.thatsbooks.com
印　　刷： 北京市房山腾龙印刷厂
版　　次： 2023 年 5 月第 1 版第 2 次印刷
I S B N： 978-7-5085-4336-9
开　　本： 787mm×1092mm　1/32
印　　张： 6
字　　数： 20 千
定　　价： 49.8 元